Dyson Sphere Chronicles

Full details of the potential concept to discover Alien life forms and impact on stellar system

By

Barbs Walters

Copyright © 2024, by Barbs Walters.

All rights reserved. No part of this book may be reproduced or transmitted in any form or by any means, electronic or mechanical, including photocopying, recording, or any information storage and retrieval system, without permission in writing from the copyright owner, except for brief quotations in critical reviews and articles.

Table of Contents

Copyright © 2024, by Barbs Walters.   2
Table of Contents   3
Introduction   5
Chapter 1: The Visionary Concept of Dyson Spheres 7
    1.1 Freeman Dyson's Revolutionary Idea   7
    1.2 Types of Dyson Structures   10
Chapter 2: The Scientific Basis for Dyson Spheres 18
    2.1 Energy Needs of Advanced Civilizations   18
    2.2 Infrared Radiation as a Detection Method   22
Chapter 3: Historical Context and Early Research   30
    3.1 Freeman Dyson's Influence and Legacy   30
    3.2 Initial Searches and Challenges   35
Chapter 4: Recent Advances in Dyson Sphere Detection   42
    4.1 Breakthrough Studies and Technologies   42
    4.2 The 2024 Study and its Findings   46
Chapter 5: Alternative Explanations for Infrared Anomalies   54
    5.1 Natural Phenomena and Astronomical Events   55
    5.2 Astrophysical Contaminants   58
Chapter 6: Future Research Directions   64
    6.1 Advanced Observational Tools   64
    6.2 Collaboration and Cross-Disciplinary

Approaches	69
Chapter 7: The Broader Implications of Dyson Spheres	**76**
7.1 Philosophical and Ethical Considerations	76
7.2 The Search for Extraterrestrial Intelligence (SETI)	81
Conclusion	**87**

Introduction

The concept of Dyson spheres, proposed by physicist Freeman Dyson in 1960, offers a fascinating glimpse into the possibilities of advanced civilizations. This book explores the scientific, technological, and philosophical dimensions of Dyson spheres, examining the implications of harnessing star energy and the potential existence of extraterrestrial civilizations capable of such feats.

Through a blend of historical context, recent research, and alternative explanations, we delve into the search for Dyson spheres and the broader impact on our understanding of life in the universe. We discuss the theoretical basis behind these advanced structures, the scientific methods used to detect them, and the challenges that remain in validating these claims.

In addition, we explore the philosophical and ethical considerations of such advanced technology, as well as the impact of discovering extraterrestrial intelligence. The implications extend to our own future, both in terms of energy solutions and space exploration. By integrating data from various disciplines, we can gain a deeper understanding of the potential and limitations of these extraordinary constructs.

Chapter 1: The Visionary Concept of Dyson Spheres

1.1 Freeman Dyson's Revolutionary Idea

Origin and Inspiration: Olaf Stapledon's "Star Maker"

The concept of a Dyson sphere has fascinated scientists and science fiction enthusiasts for decades, thanks to the visionary ideas of Freeman Dyson. To understand the origins of this concept, we need to journey back to the literary world of the 1930s. In 1937, British author Olaf Stapledon published a novel titled "Star Maker." This book, although a work of fiction, laid the groundwork for what would later become a serious scientific hypothesis.

"Star Maker" explores the evolution of life and intelligence in the universe. Stapledon's expansive narrative includes the idea of advanced civilizations harnessing the energy of entire stars. In his novel, these civilizations build colossal structures around their suns to capture and utilize stellar energy. This imaginative leap planted the seed that would later bloom in Freeman Dyson's scientific mind.

Freeman Dyson, a British-American physicist and mathematician, was deeply influenced by Stapledon's work. He saw potential in the idea of a civilization capturing the full energy output of its star. In 1960, Dyson translated this imaginative concept into a scientific proposal, suggesting that such structures could be detectable by their infrared radiation. This proposal was a leap forward, blending science fiction with real scientific inquiry.

Dyson's 1960 Paper: Defining the Dyson Sphere

In his seminal 1960 paper, "Search for Artificial Stellar Sources of Infrared Radiation," Dyson laid out his vision for what is now known as the Dyson sphere. His hypothesis was straightforward yet profound: an advanced civilization would eventually require energy on a scale that only a star could provide. To meet these needs, such a civilization might construct an enormous structure around their star to capture its energy output.

Dyson proposed that if such structures existed, they would emit waste heat in the form of infrared radiation. This radiation, he argued, could be detectable with the right instruments. Thus, by searching for unusual infrared signatures, scientists might identify signs of advanced extraterrestrial civilizations.

What set Dyson's idea apart was its practical approach to the search for extraterrestrial intelligence (SETI). Rather than looking for radio signals, as many contemporary SETI efforts did, Dyson suggested looking for physical megastructures. His paper laid the foundation for a new method of searching for alien life, one that focused on the byproducts of advanced technology.

1.2 Types of Dyson Structures

Dyson Swarm: Feasible and Practical

When Dyson first described his concept, he didn't envision a solid shell surrounding a star. Instead, he proposed a swarm of individual satellites or solar collectors orbiting the star in dense formations. This concept, known as a Dyson swarm, is considered the most feasible and practical variant of a Dyson sphere.

A Dyson swarm would consist of millions, possibly billions, of independent units. Each unit would be a solar collector or habitat, capturing solar energy and converting it into usable power. These units would orbit the star in a complex yet stable pattern, ensuring that the swarm as a whole could harness the star's energy efficiently.

The feasibility of a Dyson swarm lies in its modularity and scalability. Unlike a single monolithic structure, a swarm can be built incrementally. An advanced civilization could start with a few solar collectors and gradually expand the swarm over time. This incremental approach makes the concept more achievable, even if the initial technological and material requirements are high.

Moreover, a Dyson swarm offers flexibility. Individual units can be moved, replaced, or upgraded as needed. This adaptability is crucial for maintaining such a vast and complex system. Additionally, the swarm configuration minimizes the risk of catastrophic failure. If one unit fails, the rest of the swarm continues to function.

Building a Dyson swarm would require significant advancements in materials science, robotics, and energy storage. The logistics of constructing and maintaining billions of units in orbit around a star are daunting. However, the potential benefits—essentially limitless energy—make the concept highly appealing.

Dyson Shell: The Classic but Impractical Vision

While the Dyson swarm is the most practical implementation of Dyson's idea, the classic vision of a Dyson sphere is that of a solid shell completely enclosing a star. This Dyson shell is a vast, continuous structure capturing every bit of energy emitted by the star. Though it captures the imagination, the Dyson shell is widely regarded as impractical with current or foreseeable technology.

The challenges of building a Dyson shell are immense. First, there is the issue of materials. To construct a shell of this magnitude, an advanced civilization would need materials far stronger and more durable than anything currently available. The structure would have to withstand immense gravitational forces, radiation, and thermal stress. Dyson himself suggested that a civilization might need to dismantle entire planets to gather the necessary resources, a task of unimaginable scale.

Second, there is the problem of stability. A rigid shell would need to be positioned at a precise distance from the star to avoid being pulled in by gravity or pushed away by solar radiation pressure. Maintaining this balance would be extraordinarily complex. Any perturbation could result in catastrophic failure, either collapsing the shell into the star or tearing it apart.

Third, the energy distribution and heat management within the shell pose significant challenges. A solid shell would trap heat, creating a potentially uninhabitable environment. Advanced heat dissipation technologies would be required to manage the thermal load, ensuring that the interior remains stable and livable.

Despite these challenges, the Dyson shell remains a powerful symbol of technological achievement and ambition. It represents the pinnacle of energy capture and utilization, a testament to what a truly advanced civilization might achieve. While building a Dyson shell may be beyond our current capabilities, the concept continues to inspire scientists and futurists to push the boundaries of what is possible.

Freeman Dyson's concept of harnessing a star's energy through vast megastructures has profoundly influenced both scientific thought and popular culture. From its origins in Olaf Stapledon's "Star Maker" to Dyson's groundbreaking 1960 paper, the idea of a Dyson sphere challenges us to envision the possibilities of advanced technology and the future of energy utilization.

The practical Dyson swarm offers a feasible path toward capturing stellar energy, while the classic Dyson shell, despite its impracticality, stands as a monument to human ambition and ingenuity. As we continue to explore the cosmos and develop our technological capabilities, Dyson's visionary concept remains a guiding light, encouraging us to reach for the stars—quite literally.

In the following chapters, we will delve deeper into the scientific basis for Dyson spheres, the historical context of their research, recent advancements in detection methods, alternative explanations for observed phenomena, and the broader implications of these hypothetical structures for humanity and the search for extraterrestrial intelligence. Through this exploration, we hope to gain a better understanding of the potential and challenges of one of the most fascinating ideas in modern science.

Chapter 2: The Scientific Basis for Dyson Spheres

2.1 Energy Needs of Advanced Civilizations

The Kardashev Scale: Classifying Civilizations

To understand the concept of Dyson spheres, it's essential to grasp why an advanced civilization would need such a vast amount of energy. One way to classify the energy consumption of civilizations is through the Kardashev Scale, introduced by Soviet astronomer Nikolai Kardashev in 1964. The Kardashev Scale categorizes civilizations based on their ability to harness energy:

- Type I Civilization: This is a planetary civilization that can use and store all of the energy available on its home planet. This includes all the energy from natural resources like fossil fuels, wind, and solar power. By most estimates, humanity is currently around 0.7 on this scale, as we haven't yet fully harnessed all the energy resources available on Earth.

- Type II Civilization: A stellar civilization that can capture and use the total energy output of its star. This is where the concept of Dyson spheres comes into play. To meet the energy demands of a Type II civilization, harnessing the power of their star would be essential. The energy output of a star like our Sun is vastly greater than the total energy available on Earth, providing a nearly unlimited power source.

- Type III Civilization: A galactic civilization that can control energy on the scale of its entire galaxy. This would involve harnessing the energy from billions of stars. While this level of civilization is purely theoretical, it serves to illustrate the upper limits of energy consumption and technological advancement.

The Kardashev Scale provides a framework for understanding the progression of civilizations in terms of energy usage. As a civilization advances, its energy needs grow exponentially. For a Type II civilization, harnessing the energy of an entire star becomes not just a possibility but a necessity to sustain its technological and societal growth.

Harnessing Star Power: Meeting Vast Energy Demands

For a civilization to reach Type II status, it must find ways to meet its enormous energy demands. The energy output of a typical star like the Sun is about 3.8×10^{26} watts. To put that into perspective, this is more than 10 billion times the current energy consumption of all humanity.

Harnessing this energy would allow a civilization to power advanced technologies, support large populations, and undertake ambitious projects like interstellar travel. The concept of a Dyson sphere, or more practically a Dyson swarm, offers a theoretical solution to this challenge. By surrounding a star with a vast array of energy-collecting satellites or structures, a civilization could capture a significant portion of the star's output.

This energy could then be used to power various aspects of the civilization, from basic needs like heating and cooling to advanced applications like powering spacecraft, supporting artificial environments, and running complex simulations. The ability to harness star power represents a critical step in the evolution of an advanced civilization, providing the means to achieve technological feats that are currently beyond our imagination.

2.2 Infrared Radiation as a Detection Method

Waste Heat Emission: The Key Signature

One of the most intriguing aspects of Dyson spheres is the potential to detect them through their waste heat emissions. When an object absorbs energy, it must radiate away excess heat to maintain thermal equilibrium. This is true for all physical systems, from small electronic devices to massive stellar structures.

A Dyson sphere or swarm, by capturing a star's energy, would inevitably re-emit some of this energy as waste heat. According to the principles of thermodynamics, this waste heat would be emitted in the form of infrared radiation. Unlike visible light, which is the primary emission from stars, infrared radiation is longer-wavelength and is not visible to the naked eye. However, it can be detected by specialized telescopes and instruments.

The idea, then, is to look for stars that exhibit unusual infrared signatures. A star surrounded by a Dyson sphere would have less visible light escaping, as much of it would be absorbed and re-emitted as infrared radiation. By identifying stars with an excess of infrared emission relative to their visible light, astronomers can pinpoint candidates for further study. This method leverages the predictable physical behavior of large structures and the unique spectral signature of waste heat.

Observational Techniques: Detecting Infrared Anomalies

Detecting the faint infrared signals from potential Dyson spheres requires advanced observational techniques and sophisticated instruments. Over the years, several telescopes and surveys have been developed to explore the universe in infrared wavelengths.

One of the key tools in this search has been the Wide-field Infrared Survey Explorer (WISE), a

NASA space telescope launched in 2009. WISE scans the sky in infrared wavelengths, capturing detailed images that reveal the heat signatures of celestial objects. By analyzing data from WISE, astronomers can identify stars with unusual infrared excesses that might indicate the presence of a Dyson sphere.

Another significant instrument is the European Space Agency's Gaia observatory. While Gaia primarily maps the positions and motions of stars, its data can also be cross-referenced with infrared observations to identify anomalies. Combining data from multiple sources increases the accuracy and reliability of identifying potential Dyson sphere candidates.

When a star with an unusual infrared signature is found, astronomers perform follow-up observations using more powerful telescopes like the James Webb Space Telescope (JWST). JWST, with its advanced infrared capabilities, can provide more detailed information about the nature of these anomalies. It can help determine whether the infrared excess is due to a Dyson sphere or other natural phenomena such as dust clouds or young stellar objects.

The search for infrared anomalies is not without challenges. One major difficulty is distinguishing between different sources of infrared radiation. For example, dust clouds around young stars can emit significant infrared radiation, mimicking the signature of a Dyson sphere. Therefore, careful analysis and elimination of other possible explanations are crucial before considering an artificial origin.

Advanced data processing techniques, including machine learning and statistical analysis, are employed to sift through vast amounts of observational data. These techniques help identify patterns and anomalies that might otherwise go unnoticed. By continually refining these methods, astronomers improve their chances of detecting genuine Dyson sphere candidates.

The search for Dyson spheres through infrared detection represents a fascinating intersection of theoretical science and practical astronomy. It challenges us to develop new technologies and methodologies while expanding our understanding of the universe. Although we have yet to confirm the existence of any Dyson spheres, the pursuit itself drives innovation and deepens our appreciation of the cosmos.

The scientific basis for Dyson spheres is rooted in our understanding of energy needs for advanced civilizations and the physical principles governing waste heat emission. The Kardashev Scale provides a framework for classifying civilizations based on their energy consumption, highlighting the necessity for harnessing stellar power at advanced stages. Dyson spheres, whether in the form of swarms or shells, offer a theoretical solution to these immense energy demands.

Detecting Dyson spheres through their infrared signatures is a practical approach that leverages our knowledge of thermodynamics and advanced observational techniques. By searching for infrared anomalies, astronomers hope to uncover evidence of advanced extraterrestrial civilizations. This method not only broadens the scope of SETI but also pushes the boundaries of our technological capabilities.

As we continue to explore the universe and refine our detection methods, the search for Dyson spheres remains a symbol of human curiosity and ambition. It encourages us to dream big and reach for the stars, embodying the spirit of exploration and discovery that drives scientific progress.

Chapter 3: Historical Context and Early Research

3.1 Freeman Dyson's Influence and Legacy

Early Receptions and Criticisms

Freeman Dyson's proposal of the Dyson sphere in 1960 was a groundbreaking idea that captured the imagination of scientists and the public alike. Dyson suggested that advanced civilizations might build massive structures to harness the energy of their stars, thus solving their energy needs. The concept, while rooted in science fiction, sparked significant interest in both the scientific community and popular culture.

When Dyson first introduced the idea, the scientific community responded with a mix of curiosity and skepticism. Many scientists were intrigued by the boldness of the concept and its implications for our understanding of advanced civilizations. However, others were more cautious, pointing out the immense engineering challenges and the speculative nature of the idea.

The primary criticism centered on the feasibility of constructing such a colossal structure. A Dyson sphere would require an unimaginable amount of materials and energy to build. Critics argued that even if an advanced civilization could harness the energy of an entire star, the logistical and technological hurdles would be enormous. Additionally, the concept seemed to stretch the limits of our understanding of physics and engineering at the time.

Despite these criticisms, Dyson's idea gained traction. It pushed scientists to think about energy consumption and technological advancement on a cosmic scale. The concept also raised intriguing questions about the potential for extraterrestrial civilizations and how we might detect them. By considering the possibility of Dyson spheres, Dyson encouraged scientists to explore new avenues of thought and research.

Impact on Scientific Thought and Popular Culture

Freeman Dyson's concept had a profound impact on both scientific thought and popular culture. In the scientific realm, it encouraged researchers to think beyond the boundaries of Earth and consider the broader implications of technological advancement. The idea of Dyson spheres became a staple in discussions about the future of human civilization and the search for extraterrestrial intelligence.

Dyson's proposal also influenced the development of the Kardashev Scale, which classifies civilizations based on their energy consumption. The scale provided a framework for understanding how civilizations might progress and highlighted the importance of energy in technological development. The concept of Dyson spheres became closely associated with Type II civilizations on this scale, further cementing its place in scientific thought.

In popular culture, the idea of Dyson spheres captured the public's imagination. It appeared in numerous science fiction novels, movies, and television shows. For instance, the concept featured prominently in Larry Niven's 1970 novel "Ringworld," which imagined a ring-like structure encircling a star. The idea also appeared in episodes of "Star Trek" and other science fiction series, where it served as a symbol of advanced technology and the potential of human ingenuity.

The influence of Dyson spheres extended beyond science fiction. The concept inspired artists, writers, and thinkers to explore themes of energy, technology, and the future of civilization. It encouraged people to dream about what humanity could achieve if we could harness the power of the stars. Dyson's idea became a symbol of human aspiration and the pursuit of knowledge.

3.2 Initial Searches and Challenges

SETI and Infrared Astronomy: Early Efforts

The search for Dyson spheres became a key aspect of the Search for Extraterrestrial Intelligence (SETI). In the 1960s, SETI researchers began considering the possibility that advanced civilizations might leave detectable signatures in the form of waste heat. Dyson's idea that Dyson spheres would emit infrared radiation provided a potential method for detecting such civilizations.

Early efforts to search for Dyson spheres focused on infrared astronomy. Infrared radiation is a type of electromagnetic radiation with longer wavelengths than visible light. It is often associated with heat, making it a prime candidate for detecting the waste heat emitted by a Dyson sphere. Scientists began looking for stars with unusual infrared signatures that could indicate the presence of such structures.

One of the first major infrared surveys was the Infrared Astronomical Satellite (IRAS) mission, launched in 1983. IRAS conducted an all-sky survey, capturing infrared data from numerous stars and galaxies. Although the primary goal of IRAS was not to search for Dyson spheres, the data it collected provided valuable information for SETI researchers. By analyzing this data, scientists hoped to identify stars with excess infrared emission that could not be explained by natural phenomena.

Despite the potential of infrared astronomy, early searches for Dyson spheres faced significant challenges. The sensitivity of infrared detectors in the 1960s and 1980s was limited, making it difficult to detect faint infrared signatures. Additionally, many natural sources of infrared radiation, such as dust clouds and young stars, could mimic the signature of a Dyson sphere. This made it challenging to distinguish between potential Dyson spheres and natural astronomical objects.

Technological Limitations of the 1960s

The technological limitations of the 1960s posed significant hurdles for early searches for

Dyson spheres. Infrared detectors were relatively primitive by today's standards, with limited sensitivity and resolution. This made it difficult to identify faint infrared signals that might indicate the presence of a Dyson sphere.

Moreover, the computing power available at the time was insufficient for processing large amounts of infrared data. Early computers lacked the speed and storage capacity needed to analyze the vast datasets generated by infrared surveys. This limited the ability of researchers to search for subtle anomalies in the data that could point to Dyson spheres.

Another challenge was the lack of comprehensive infrared surveys. While IRAS provided valuable data in the 1980s, there were no equivalent surveys in the 1960s. This meant that early searches had to rely on limited and often incomplete data. The absence of large-scale infrared surveys hindered the ability of researchers to conduct thorough searches for Dyson spheres.

Despite these challenges, early researchers made significant strides in developing the methods and technologies needed to search for Dyson spheres. They laid the groundwork for future searches by establishing the importance of infrared astronomy and developing techniques for analyzing infrared data. These early efforts demonstrated the potential of infrared detection methods and encouraged further research in this area.

Freeman Dyson's proposal of Dyson spheres had a lasting impact on both scientific thought and popular culture. While the concept faced early criticism, it pushed scientists to consider new ideas about energy consumption and technological advancement. The influence of Dyson spheres extended beyond the scientific community, inspiring writers, artists, and thinkers to explore themes of energy and the future of civilization.

Early searches for Dyson spheres, conducted as part of SETI, focused on detecting infrared radiation. These efforts faced significant challenges due to the technological limitations of the 1960s and the complexities of infrared astronomy. Despite these obstacles, early researchers made important contributions to the development of methods for detecting Dyson spheres and laid the groundwork for future searches.

The search for Dyson spheres continues to this day, driven by advances in infrared technology and data analysis. By building on the foundation established by early researchers, modern scientists are better equipped to explore the possibility of advanced extraterrestrial civilizations and the role of Dyson spheres in meeting their energy needs. The journey to understand and detect Dyson spheres remains a testament to human curiosity and the relentless pursuit of knowledge.

Chapter 4: Recent Advances in Dyson Sphere Detection

4.1 Breakthrough Studies and Technologies

Advances in Infrared Telescopes

In recent years, the search for Dyson spheres has been greatly enhanced by significant advancements in infrared telescope technology. Infrared telescopes are designed to detect infrared radiation, which is essentially heat emitted by objects. For a long time, the sensitivity and resolution of infrared detectors were limited, making it difficult to identify the faint signals that might indicate the presence of a Dyson sphere. However, technological progress has revolutionized this field.

Modern infrared telescopes are far more sensitive and capable than their predecessors. They can detect minute amounts of infrared radiation with high precision, allowing astronomers to observe distant stars and galaxies in unprecedented detail. These telescopes can also cover a broader range of wavelengths, capturing more comprehensive data that helps distinguish between different sources of infrared radiation.

One of the most significant breakthroughs came with the development of space-based infrared telescopes. Unlike ground-based telescopes, which are affected by the Earth's atmosphere, space-based telescopes can observe the cosmos without interference. This results in much clearer and more accurate data. The deployment of telescopes like NASA's Wide-field Infrared Survey Explorer (WISE) and the European Space Agency's Gaia observatory has been crucial in advancing our understanding of the universe.

Contributions from NASA's WISE and ESA's Gaia

NASA's WISE telescope, launched in 2009, was a game-changer for infrared astronomy. WISE conducted an all-sky survey, capturing images of the entire sky in infrared light. This mission provided astronomers with a treasure trove of data, revealing countless stars, galaxies, and other celestial objects. The detailed infrared maps created by WISE have been instrumental in the search for Dyson spheres.

WISE's ability to detect faint infrared signals allowed researchers to look for anomalies that could indicate the presence of a Dyson sphere. By analyzing the data from WISE, scientists could identify stars that emitted more infrared radiation than expected. These anomalies are essential clues in the search for advanced extraterrestrial civilizations that might be harnessing the energy of their stars.

Similarly, the European Space Agency's Gaia observatory, launched in 2013, has made significant contributions to this field. While Gaia's primary mission is to create a detailed 3D map of the Milky Way, it also collects valuable data on the brightness and position of stars. Gaia's precise measurements help astronomers understand the characteristics of stars, including those with unusual infrared emissions.

The combination of data from WISE and Gaia has provided a more comprehensive view of the stars in our galaxy. By cross-referencing information from both telescopes, researchers can identify stars with unexplained infrared radiation more accurately. This collaborative approach enhances the reliability of potential Dyson sphere candidates and reduces the likelihood of false positives.

4.2 The 2024 Study and its Findings

Methodology: Analyzing 5 Million Stars

A recent study published in 2024 represents a major milestone in the search for Dyson spheres. The study focused on analyzing data from 5 million stars within 1,000 light-years of Earth, using information from both the WISE and Gaia telescopes. The goal was to identify stars with infrared signatures that could not be explained by natural phenomena.

The researchers employed a rigorous methodology to sift through the massive dataset. They started by applying filters to remove data contamination and isolate stars with significant infrared emissions. This initial filtering process reduced the sample size considerably, allowing the researchers to focus on the most promising candidates.

Next, the team conducted a detailed analysis of the remaining stars. They examined various factors that could cause excess infrared radiation, such as interstellar dust, young stars with debris disks, and planetary collisions. By ruling out these natural explanations, the researchers aimed to identify stars with infrared signatures that matched the theoretical predictions for Dyson spheres.

One of the key challenges was distinguishing between potential Dyson spheres and natural objects. To address this, the researchers developed sophisticated models to simulate the infrared emissions of Dyson spheres. These models accounted for various configurations, such as Dyson swarms and Dyson shells, and helped the team compare the observed data with theoretical predictions.

Seven Promising Candidates: Detailed Analysis

The 2024 study ultimately identified seven stars with unexplained infrared emissions, suggesting they could be potential Dyson sphere candidates. These stars stood out because their infrared signatures did not match any known natural phenomena. While this does not confirm the existence of Dyson spheres, it provides intriguing evidence that warrants further investigation.

The seven candidate stars are all red dwarfs, the most common type of star in our galaxy. Red dwarfs are smaller and cooler than our Sun, making their infrared emissions easier to detect. The researchers found that these stars emitted more infrared radiation than expected, consistent with the presence of structures like Dyson spheres.

Each of the seven candidates underwent a thorough analysis to rule out other possible explanations. For example, the team considered the possibility that the infrared emissions could be caused by a galaxy in the background, creating an overlapping signal. They also looked into the likelihood of planetary collisions or debris disks around the stars. However, none of these natural causes could fully explain the observed infrared signatures.

The researchers used data from both WISE and Gaia to cross-verify their findings. By combining information from two different sources, they increased the accuracy and reliability of their results. This approach minimized the risk of false positives and strengthened the case for these stars being genuine Dyson sphere candidates.

The study's findings are significant because they provide a new direction for the search for extraterrestrial intelligence. The seven candidate stars represent a starting point for further research and observation. Future missions, such as the James Webb Space Telescope, could provide more detailed data on these stars and help determine whether they are indeed surrounded by Dyson spheres.

The identification of these candidates also highlights the importance of continued advancements in infrared astronomy. As technology improves, so does our ability to detect and analyze faint infrared signals from distant stars. The 2024 study demonstrates the potential of modern telescopes and data analysis techniques to uncover evidence of advanced civilizations.

The search for Dyson spheres has come a long way since Freeman Dyson first proposed the concept in 1960. Advances in infrared telescope technology, particularly the contributions from NASA's WISE and ESA's Gaia, have significantly enhanced our ability to detect faint infrared emissions from distant stars. These advancements have paved the way for groundbreaking studies, such as the 2024 study that identified seven potential Dyson sphere candidates.

The methodology used in the 2024 study, which involved analyzing data from 5 million stars, represents a major leap forward in the search for advanced extraterrestrial civilizations. By employing rigorous filtering and detailed analysis techniques, the researchers were able to identify stars with unexplained infrared emissions that match the theoretical predictions for Dyson spheres.

While the discovery of these seven candidates does not confirm the existence of Dyson spheres, it provides compelling evidence that warrants further investigation. The findings of the 2024 study highlight the importance of continued research and technological advancements in the field of infrared astronomy. As we develop more powerful telescopes and data analysis methods, we inch closer to uncovering the mysteries of the universe and the potential existence of advanced civilizations beyond our own.

Chapter 5: Alternative Explanations for Infrared Anomalies

The search for Dyson spheres is a fascinating endeavor, driven by the hope of finding evidence of advanced extraterrestrial civilizations. However, it is crucial to approach the data with a healthy dose of skepticism. Many natural phenomena and astronomical events can produce infrared anomalies that might be mistaken for signs of advanced technology. This chapter delves into these alternative explanations, focusing on natural processes and astrophysical contaminants that can create similar infrared signatures.

5.1 Natural Phenomena and Astronomical Events

Planetary Collisions: Debris Disks and Their Signatures

One of the most compelling natural explanations for unusual infrared emissions is planetary collisions. When planets collide, they create a massive amount of debris that forms a disk around the star. This debris disk can emit significant amounts of infrared radiation, which might be mistaken for the waste heat of a Dyson sphere.

Planetary collisions are not uncommon in the universe. In the early stages of a star system's formation, planets and other celestial bodies frequently collide as they settle into stable orbits. Even in more mature systems, gravitational interactions can lead to catastrophic collisions. These events generate debris disks that are heated by the star, causing them to glow in infrared light.

Debris disks are often temporary phenomena. Over time, the debris either falls into the star or coalesces into larger bodies, such as planets or moons. However, while they exist, these disks can produce infrared emissions strong enough to mimic the signature of a Dyson sphere. Therefore, when researchers detect an infrared anomaly, they must consider the possibility of a recent planetary collision.

Young Stars: Infrared Emissions from Hot Debris

Young stars often have high levels of infrared emissions due to the presence of hot debris in their surrounding environment. When a star forms, it is surrounded by a protoplanetary disk composed of gas, dust, and other materials. As planets begin to form within this disk, the collisions and interactions between these nascent bodies generate significant amounts of heat.

This hot debris can emit infrared radiation that might be mistaken for a Dyson sphere. Young stars, typically classified as T Tauri stars or Herbig Ae/Be stars, are often surrounded by these protoplanetary disks. The infrared emissions from these disks can be quite strong, leading to potential false positives in the search for Dyson spheres.

Furthermore, the infrared signature of a young star with a protoplanetary disk can vary over time. As planets form and the disk evolves, the levels of infrared radiation can change, complicating the task of distinguishing between natural phenomena and artificial structures. Scientists must carefully analyze the age and characteristics of stars with infrared anomalies to determine whether they are young stars with hot debris or potential candidates for Dyson spheres.

5.2 Astrophysical Contaminants

Background Galaxies: Misinterpretation Risks

Another major challenge in the search for Dyson spheres is the potential for misinterpreting the infrared emissions from background galaxies. When observing distant stars, it is possible that a galaxy in the background might overlap with the star being studied. This overlap can cause the combined infrared emissions of the star and the galaxy to appear as an anomaly.

Background galaxies can emit significant amounts of infrared radiation, especially if they are active or contain large amounts of dust. These emissions can easily be mistaken for the waste heat of a Dyson sphere. The risk of misinterpretation is particularly high when the background galaxy is faint and not immediately visible in optical observations, making it difficult to distinguish its contribution to the infrared signal.

To mitigate this risk, astronomers use multiple wavelengths and observational techniques to separate the contributions of foreground stars and background galaxies. Cross-referencing data from different telescopes and surveys can help identify and account for background galaxies. However, the possibility of misinterpretation always exists, and researchers must remain vigilant in their analysis.

Hot Dust-Obscured Galaxies (Hot DOGs)

Hot Dust-Obscured Galaxies, or Hot DOGs, are a specific type of galaxy that can complicate the search for Dyson spheres. These galaxies are characterized by their extreme levels of infrared emissions, caused by large amounts of hot dust that obscure their light in other wavelengths. Hot DOGs are rare but extremely bright in the infrared spectrum, making them potential sources of false positives.

Hot DOGs were first identified in data from the WISE telescope, which was designed to detect infrared radiation. These galaxies are often undergoing intense star formation or have active galactic nuclei (AGN) that heat the surrounding dust. The resulting infrared emissions can be so strong that they outshine other sources in the same field of view.

When analyzing infrared data, astronomers must consider the possibility that an observed anomaly could be a Hot DOG rather than a Dyson sphere. This requires careful cross-referencing with other astronomical data, such as optical and radio observations, to identify and rule out the presence of such galaxies. The study of Hot DOGs has become an important aspect of understanding the broader context of infrared emissions in the universe.

The search for Dyson spheres is an exciting and challenging endeavor. While the potential discovery of an advanced extraterrestrial civilization would be groundbreaking, it is essential to consider and rule out natural explanations for infrared anomalies. Planetary collisions, young stars with hot debris, background galaxies, and Hot Dust-Obscured Galaxies all present plausible alternative explanations for unusual infrared emissions.

By understanding these natural phenomena and astrophysical contaminants, scientists can refine their search methods and improve the accuracy of their detections. The use of advanced infrared telescopes and multi-wavelength observations is crucial in distinguishing between natural and artificial sources of infrared radiation. As our technology and techniques continue to improve, the dream of discovering evidence of advanced civilizations remains a tantalizing possibility, but one that requires careful and rigorous scientific investigation.

Chapter 6: Future Research Directions

As the search for Dyson spheres and other signs of advanced extraterrestrial civilizations continues, future research will depend on advanced observational tools and interdisciplinary collaboration. This chapter explores the promising technologies and research methodologies that could revolutionize our understanding of the universe and potentially lead to groundbreaking discoveries.

6.1 Advanced Observational Tools

The James Webb Space Telescope: Capabilities and Potential

The James Webb Space Telescope (JWST) represents a significant leap forward in our ability to observe the universe. Launched by NASA in December 2021, JWST is the most powerful space telescope ever built, designed to look deeper into space and time than its predecessors. Its advanced capabilities make it a crucial tool in the search for Dyson spheres.

JWST's primary mirror is 6.5 meters in diameter, much larger than the Hubble Space Telescope's 2.4-meter mirror. This larger mirror allows JWST to collect more light, enabling it to observe faint objects with unprecedented clarity. The telescope is optimized for infrared observations, which is essential for detecting the waste heat emissions of potential Dyson spheres.

One of JWST's key advantages is its ability to observe in the mid-infrared range, where the waste heat from a Dyson sphere would be most apparent. Its instruments can detect the faint infrared signatures that might indicate the presence of such structures around distant stars. By analyzing the spectra of these emissions, scientists can distinguish between natural sources of infrared radiation and potential signs of advanced technology.

In addition to its infrared capabilities, JWST's ability to observe exoplanet atmospheres could provide indirect evidence of advanced civilizations. By studying the chemical composition of exoplanet atmospheres, researchers can look for signs of industrial pollutants or other markers of technological activity. This dual capability makes JWST an invaluable asset in the search for extraterrestrial intelligence.

Planning Future Studies: Securing Telescope Time

Securing observation time on the James Webb Space Telescope is a competitive process, as it is one of the most sought-after instruments in the field of astronomy. Researchers must submit detailed proposals outlining their scientific goals and the significance of their work. These proposals are reviewed by committees of experts who allocate telescope time based on scientific merit and potential impact.

Given the importance of detecting Dyson spheres, proposals related to this search are likely to receive serious consideration. However, researchers must ensure their proposals are robust and well-justified. This involves demonstrating a clear methodology for identifying and analyzing infrared anomalies and showing how JWST's capabilities will be utilized to achieve their objectives.

In addition to JWST, other advanced telescopes, such as the Extremely Large Telescope (ELT) currently under construction in Chile, will provide complementary capabilities. The ELT's massive 39-meter mirror will offer unparalleled resolution and sensitivity, allowing for detailed observations of distant stars and potential Dyson sphere candidates. Combining data from JWST and ELT will enable a more comprehensive analysis of infrared anomalies and enhance the likelihood of identifying signs of advanced civilizations.

6.2 Collaboration and Cross-Disciplinary Approaches

Integrating Data from Multiple Surveys

One of the keys to advancing the search for Dyson spheres is integrating data from multiple astronomical surveys. Different telescopes and instruments provide complementary information, and combining these data sets can lead to more accurate and comprehensive analyses.

For example, data from the Wide-field Infrared Survey Explorer (WISE), which conducted an all-sky survey in infrared light, can be combined with observations from the Gaia mission, which maps the positions and motions of stars with high precision. WISE data can identify infrared anomalies, while Gaia data can provide context about the stars exhibiting these anomalies, such as their distances, ages, and compositions.

Combining data from different wavelengths is also crucial. Infrared observations can reveal the heat signatures of potential Dyson spheres, while optical and ultraviolet observations can provide additional information about the stars and any surrounding planets or debris. Radio observations can detect other signs of technological activity, such as artificial signals or unusual radio emissions.

Advanced data integration techniques, such as machine learning and artificial intelligence, can help process and analyze these large and complex data sets. By training algorithms to recognize patterns associated with Dyson spheres, researchers can more efficiently identify promising candidates and rule out false positives. These techniques can also help prioritize targets for follow-up observations with more powerful telescopes like JWST and ELT.

Interdisciplinary Research: Physics, Astronomy, and Beyond

The search for Dyson spheres is inherently interdisciplinary, requiring collaboration between experts in various fields of science. Astronomers, physicists, and engineers must work together to develop and refine the observational techniques needed to detect these structures. Theoretical physicists can help model the properties and behaviors of Dyson spheres, providing a framework for interpreting observational data.

In addition to the physical sciences, the search for extraterrestrial intelligence involves other disciplines, such as astrobiology, computer science, and even philosophy. Astrobiologists study the conditions necessary for life and how it might evolve into technologically advanced civilizations. Their insights can inform the search for Dyson spheres by identifying the types of stars and planetary systems most likely to host such civilizations.

Computer scientists play a crucial role in developing the algorithms and data analysis tools needed to process vast amounts of astronomical data. Machine learning and artificial intelligence techniques can help identify patterns and anomalies that might indicate the presence of Dyson spheres. These tools can also be used to simulate different scenarios and refine search strategies.

Philosophers and ethicists contribute by exploring the implications of discovering advanced extraterrestrial civilizations. They help frame the scientific questions and guide the interpretation of potential findings. Understanding the ethical and societal impacts of such a discovery is crucial, as it would fundamentally alter our understanding of our place in the universe.

Collaboration between these disciplines is facilitated by organizations such as the SETI Institute, which brings together scientists from diverse fields to search for extraterrestrial intelligence. Conferences, workshops, and collaborative research projects provide opportunities for experts to share knowledge, develop new methodologies, and advance the search for Dyson spheres.

The future of Dyson sphere research is bright, thanks to advancements in observational tools and the collaborative efforts of scientists across disciplines. The James Webb Space Telescope and other next-generation observatories will provide the capabilities needed to detect potential Dyson spheres and other signs of advanced civilizations. By integrating data from multiple surveys and leveraging interdisciplinary research, we can refine our search strategies and increase the chances of making groundbreaking discoveries.

As we continue to explore the universe, the search for Dyson spheres represents a quest to answer one of humanity's most profound questions: Are we alone in the universe? While the challenges are significant, the potential rewards are immense. Discovering evidence of advanced extraterrestrial civilizations would revolutionize our understanding of life, technology, and our place in the cosmos. Through careful planning, advanced technology, and collaborative research, we move closer to uncovering the mysteries of the universe and perhaps finding our cosmic neighbors.

Chapter 7: The Broader Implications of Dyson Spheres

The concept of Dyson spheres extends beyond scientific curiosity and technological possibilities; it touches on profound philosophical and ethical questions. This chapter explores the broader implications of Dyson spheres, from the potential benefits and risks of technological growth to the profound impact on humanity's understanding of life in the universe.

7.1 Philosophical and Ethical Considerations

Technological Growth: Benefits and Risks

Dyson spheres symbolize the pinnacle of technological advancement. The idea that a civilization could harness the energy output of an entire star highlights both the incredible potential and significant risks associated with such growth. On the one hand, the ability to capture vast amounts of energy could solve many of the problems we face today, such as energy scarcity and environmental degradation. On the other hand, the pursuit of such technological feats could come with unforeseen consequences.

One of the major benefits of building a Dyson sphere is the almost limitless energy supply it would provide. For an advanced civilization, this could mean the end of energy shortages, the ability to power interstellar travel, and the capacity to build massive space habitats. With energy no longer a limiting factor, humanity could focus on other challenges, such as improving quality of life and exploring the cosmos.

However, the risks are equally significant. The construction of a Dyson sphere would require dismantling entire planets, which raises questions about environmental destruction on an unprecedented scale. The ethical implications of such actions are profound. Would it be morally justifiable to destroy a planet to fuel our energy needs? This question forces us to consider the balance between technological progress and environmental stewardship.

Moreover, the drive for such technological growth could lead to social and political challenges. The resources required to build a Dyson sphere would be immense, potentially leading to conflicts over resource allocation. Additionally, the societal impacts of such advanced technology could widen the gap between different socioeconomic groups, leading to new forms of inequality.

Implications for Humanity: Energy Solutions and Space Exploration

The potential energy solutions offered by Dyson spheres could revolutionize life on Earth. Imagine a world where energy is abundant and clean, reducing our reliance on fossil fuels and mitigating the impacts of climate change. The technological advancements necessary to build a Dyson sphere could also lead to breakthroughs in other areas, such as materials science, robotics, and artificial intelligence.

The exploration of space would also take on a new dimension. With the energy provided by a Dyson sphere, humanity could undertake long-term missions to other star systems, potentially colonizing other planets and expanding our presence in the galaxy. This would not only ensure the survival of our species but also lead to the discovery of new worlds and the possibility of encountering other forms of life.

However, these possibilities come with ethical considerations. The colonization of other planets, for example, raises questions about the rights of potential extraterrestrial ecosystems. If we were to encounter microbial life on another planet, would we have the right to alter that planet's environment to suit our needs? These questions require careful consideration and international cooperation to ensure that our expansion into space is conducted responsibly and ethically.

7.2 The Search for Extraterrestrial Intelligence (SETI)

The Role of Dyson Spheres in SETI

The search for extraterrestrial intelligence (SETI) has traditionally focused on detecting artificial radio signals from other civilizations. However, Dyson spheres provide an alternative method of detection. Instead of looking for signals, we can search for the waste heat generated by a civilization that has built a Dyson sphere around its star.

Dyson spheres could serve as beacons of advanced civilizations. If we were to detect the infrared signature of a Dyson sphere, it would provide compelling evidence that we are not alone in the universe. This would be a monumental discovery, fundamentally altering our understanding of our place in the cosmos.

The role of Dyson spheres in SETI highlights the importance of considering multiple methods of detection. By broadening our search to include the detection of technological artifacts, we increase our chances of finding extraterrestrial intelligence. This approach also allows us to learn more about the potential technological capabilities of other civilizations, providing insights into our own future possibilities.

Broader Impact on the Understanding of Life in the Universe

The discovery of a Dyson sphere would have a profound impact on our understanding of life in the universe. It would confirm that intelligent life can develop the technological capability to harness the energy of an entire star. This, in turn, would suggest that other advanced civilizations might exist, each with their own unique technologies and ways of life.

Such a discovery would also have significant philosophical implications. It would challenge our understanding of what it means to be an intelligent species and force us to reconsider our assumptions about the uniqueness of human civilization. The knowledge that other intelligent beings exist and have achieved such technological feats could inspire us to strive for greater achievements and to think more broadly about our place in the cosmos.

Furthermore, the discovery of Dyson spheres would likely stimulate new scientific research and exploration. Scientists would be eager to learn more about the civilizations that built these structures, their cultures, and their histories. This could lead to a new era of space exploration, driven by the desire to learn from other intelligent species and to collaborate with them in the search for knowledge.

The impact on society would be equally profound. The knowledge that we are not alone in the universe could bring humanity together, fostering a sense of unity and common purpose. It could also lead to new ethical and philosophical debates about our responsibilities as an intelligent species and our role in the universe.

The broader implications of Dyson spheres extend far beyond the realm of science and technology. They touch on fundamental questions about our place in the universe, the ethical considerations of technological growth, and the potential for encountering other intelligent civilizations. As we continue to search for Dyson spheres and other signs of extraterrestrial intelligence, we must remain mindful of these broader implications and strive to conduct our research responsibly and ethically.

The potential benefits of discovering Dyson spheres are immense, from solving our energy needs to expanding our presence in the galaxy. However, these possibilities come with significant risks and ethical considerations that must be carefully weighed. By approaching this research with an interdisciplinary mindset and a commitment to ethical principles, we can ensure that our search for Dyson spheres contributes positively to our understanding of the universe and our place within it.

As we look to the future, the search for Dyson spheres offers a unique opportunity to explore the intersection of science, philosophy, and ethics. It challenges us to think beyond our immediate technological capabilities and consider the long-term impact of our actions on the universe. Through careful planning, advanced technology, and collaborative research, we can continue to push the boundaries of human knowledge and strive to answer one of humanity's most profound questions: Are we alone in the universe?

Conclusion

This book provides a comprehensive overview of the Dyson sphere concept, blending scientific insight with broader philosophical and ethical considerations. It emphasizes the importance of advanced technologies in our quest to understand the universe and the potential implications for humanity's future.

From the pioneering ideas of Freeman Dyson to recent advances in infrared detection techniques, the journey to uncovering these enigmatic structures is an ongoing exploration. The possibility of encountering other intelligent civilizations not only pushes the boundaries of our knowledge but also invites us to ponder our own place in the cosmos.

In conclusion, this book underscores the importance of interdisciplinary collaboration and the thoughtful examination of technological progress. It encourages readers to consider the broader ramifications of our pursuit of knowledge, fostering a deeper appreciation for the interconnectedness of science, technology, and ethics in shaping our understanding of the universe.

www.ingramcontent.com/pod-product-compliance
Lightning Source LLC
Chambersburg PA
CBHW070116230526
45472CB00004B/1291